Introduction

We no longer have black and white televisions because technology has advanced and moved forward. Pest control has not, in my humble opinion. Blasting your house with a pump sprayer was done in the 50s and is still being done today. Most companies come onto your property and spray chemicals on your baseboards. I call it the shotgun approach; they blast that all over your property regardless of the problem. As you will see in this book, that is like a black and white television. It is outdated.

Entry points should be the focus to keep the bugs out, not a toxic spray to kill them once they get inside.

Insecticides are microencapsulated, which means they are sticky, so the bugs feel something on their foot and lick it off, much like when you step on a gum. It bothers us, so we stop and get the gum off our shoes. So, when the bug guy sprays inside your house, this sticky residue gets stuck to you as you touch where he sprayed, even though it is dry. When they spray your exterior for, let's say, mosquitoes, they say don't go outside for at least one hour so the chemical can dry. When it is dry, you still pick it up when you touch anything outside, like kids' toys, patio furniture, dog bowls, and anything outside.

The chemicals they use are a neurotoxin, which attacks the bug's nervous system. When we touch whatever has been sprayed, and it is not direct spray, the drift from what was sprayed can go quite away depending on the wind that day, it also goes into our skin and affects our nervous system. Neurotoxins stored in our fat cells take their toll many years down the road.

You are paying a company to poison your family slowly.

The following book shows a different approach to staying bug-free without harsh chemicals and saves you money as well. ENJOY!

Here is a FACT if a product can cause harm to humans it is required to have a chemical label and a SDS sheet explaining the potential hazards associated with the product by the FDA (the food and Drug Administration. Products like glass cleaner and your multi-purpose cleaners have a chemical label. Products have

LD50 rating, which is the lethal dose time 50 that it takes to kill the intended insect. Products that are labeled 25B are products have been found to cause no harm to humans and do not require a chemical label or an SDS. Our company was founded on using the safest product we can find to effectively control bugs.

Let's say you find ants on your kitchen counter. What is your first reaction? Get the bug spray out and blast them. Then, you get out your spray cleaner to clean up the bug spray from the kitchen counter. Why not just eliminate step one and keep the bad bug spray off your nice clean kitchen counter where you prepare food? Simply spray the ants with, let's say, the window cleaner, and it will not only kill the ants but erase the pheromone trail they left on your counter to lead their buddies back into your kitchen. Most household cleaners will kill bugs without using the toxic bug sprays or paying a company to come into your home and spray their brand of toxic bug killer. It's safer and saves you money and time.

We will go over most of the common bugs people get in their homes and how to eliminate them without spraying toxic products or hiring someone to use those type of products.

I have been in the pest control industry for over twenty years and have run my own company since 2009. I got my start at Terminix and was hired as a termite elimination salesman. Part of the hiring process was to go to Bug school for a week and then go into a branch and learn hands on. This training required us to ride a long with a service tech for a week as his helper to see exactly what we were selling and how it was done.

We were taught in bug school to read the label before we used a product and I remember asking my trainer what the label said about using personal safety equipment like gloves and that we should be using a respirator. His response was that it was okay and not necessary. Now I saw others that were using protective gear, but more didn't than did.

I went on with my sales career with then and kind of forgot about it for a while until I started thinking about starting my own company. At that time, I decided that there must be a better and safer way to control bugs and it has led me on my journey to where I am now writing this book to educate people on safer ways to control bugs.

We will go bug by bug and discuss the why and how to eliminate them.

Gnats in your kitchen. Gnats are a bacteria problem, not a pest problem; allow me to explain. Momma gnat will lay her eggs in bacteria so her sweet little baby gnats will have food to live off until they are old enough to fly and find their own food. Ever seen the little bit of black on a banana at the end where it was torn from the tree? That is bacteria, and as with anything growing on a tree, when it is torn or removed from the tree and not sanitized, it grows bacteria, and momma gnat finds it and lays her eggs in it, and when you bring it home, the eggs begin to hatch and *boom*!! You now have gnats in your home. Once they are in your home, they look for bacteria to lay their eggs in so these new babies can have bacteria to live off until they can fly around your home and drive you crazy. Where would a gnat find bacteria in your nice clean house? Look at your garbage disposal. With the disposal off, run your finger on the bottom of that back piece of rubber and look at that nasty stuff that is now on your fingers. Clean off all that nasty stuff! Also, clean where your P traps are under your sinks. A P trap is that little curve pipe under all your sinks. It holds water and keeps all the smelly gasses in the sewer system from getting into your house. At the top of that water is a ring where bacteria form and build a nice ring inside your pipes, another egg hiding place for momma gnat. A foaming drain cleaner will eat away the bacteria and eliminate her eggs spot. The other place gnats breed in the plants from the soil or dirt in your house. There the eggs are dormant until the moisture is just right and then they hatch and will lay eggs in the nice fertile breeding ground they have found. This one is simple to solve, just give the plant a good twenty-four hours outside in the sunlight, and the UV rays from the sun will kill the bacteria and eliminate the bug issue. No Pest guy is needed.

Rodents Are Next on Our List. When you find evidence of rodent activity, it does not matter if it is a rat or a small mouse. It must go and go NOW! Most people will get some rat poison for a reason like it is quick and easy to put out, cheaper than an exterminator and no one else needs to know about it. Rat poison works and is usually within a couple of days. However, it leads to another problem. Mice do not smell as bad as a dead rat. Mice have very little bodily fluids compared to a normal rat.

THE SMELL! Rodents die in the wall, and it can smell damn bad. The old rat bait was strychnine based, and it made the rat thirsty, and it was supposed to seek water and die outside.

The other issue is what if a cat or bird eats the rodent? It will kill the eater of the poisoned rodent. A secondary kill is not good; the newer baits act completely different and will not kill whatever eats the rodents, helping to eliminate the secondary kills. It is an anticoagulant causing them to basically bleed out when a blood vessel breaks and can lead to a mess of blood... and the smell.

Now, what about traps? Snap traps baited with peanut butter work well and kill quickly versus the glue trap, where rodents get stuck and can take a long, slow death and cries can be heard inside the house. Plus, you still have the smell issue waiting to invade your home.

What about using a predator pee to make the rodents leave? NO DEATH SMELL involved with this one. Predator pee has been used for a long time to mark territory and scare off unwanted visitors. Coyote urine works well with rodents an racoon urine works on smaller rodents like mice; they smell it and want to run away from predators who can eat them for dinner. It is nothing like the death smell of a rodent body, and you spray it in the attic or garage, under the sinks, where you see the evidence of your furry friends. Now you do not need to bring a coyote home and have him pee on everything. Just order it online at predatorpee.com

But long-term best solution is to find the hole they used to gain access to your house. Exclusion work involves sealing the entry points for the little critters. We use a product called excluder cloth. It comes in a roll about three inches wide and you cut s piece and stuff the hole. It is made with stainless steel and won't cause rust stains on the building. For larger holes we use Hardware cloth with ¼ inch holes, It is similar to chicken wire, just a bit stronger and much smaller holes Exclusion is bit more work than setting traps but once it is done it is done.

Here is a bit of useless rat knowledge for you. Rats' nests are controlled by an Alpha male rat who breeds with all females in his territory, He will kill all males in his territory even his babies. For this reason, the female will leave the nest to go have her babies away from him but returns to him shortly afterwards. Rats have no bladder control, so they pee and poop as they move. And pregnant females will follow the pee trail of other females for a safe place to have babies.

Sticky Traps Are Up Next. These little dudes are pest control 24/7 and will work while you are sleeping, at work or on vacation. There are two types: the tray for big stuff or the folding paper traps for everything else. The paper one's fold into little boxes, and bugs look for a hiding spot only to be trapped and die in the trap. They are great to put near entry points like the edge of garage doors, exit doors, attics, hot water heaters, and plumbing pipes. They will and can catch anything that goes into them, like big flying American roaches, ants, silverfish, earwigs, spiders, mice, snakes, pill bugs, crickets and anything else that goes inside the trap - does not come out. They are simple to use and cheap to buy. www.thegreenbugman.com.

Sticky traps will catch everything that goes into them such as the family cat or dog. Simply use some vegetable oil or spray to break down the glue and free whoever is stuck.

Pantry Pests Are Next on The List. These range from meal moths to weevils and beetles.

Here is a fun fact for you to think about; the FDA and your local health department have acceptable amounts of bug larvae in packaged food that is sold and processed in foods we eat. The larvae are in the foods, and when they get exposed to fresh air when you open the package, they begin to hatch in your pantry or cabinets. These are foods such as flour, rice cereals, crackers, sugar, dog treats and pet foods. Spays only kill the flyers, but there are usually a lot of eggs, which can continue for a while, so the best thing to do is to clean out your pantry or cabinets, throw away the products, and get rid of the eggs. The exterminator really can't stop the problem just do a pantry cleanout.

Flies Are Next on Our List. Flies are similar to gnats and think about bacteria. Where do maggots come from? From flies, of course. Like the gnat, the momma lays her eggs in bacteria so baby flies can eat until they become fliers and drive you crazy. Trash cans are the number one source for house flies.

The very large flies are blowflies, and there is something dead around they are attracted to. Web out is a natural product that will repel them, and so will class cleaner. Removing bacteria is key for flies and gnats.

Let's Talk About Entry Points Next. How the hell do pests get in your house anyway?

Plumbing pipes; look under all your sinks and where the pipes come out from the wall. There is probably a nice plastic or chrome cover around the pipe; slide it forward just a bit and see the large hole the cover is hiding, and the drain pipes as well. Some builders seal the hole before adding covers, but most do not. Also, where the pipe comes out from the wall to fill the toilet will be the same. Sealing these pipe holes is one of the best ways to permanently keep out the pests. This applies to slab homes and pier and beam houses just the same.

The Biggie is Next. Weep holes are the number one entry point in your brick home.

Where the first row of bricks meets the slab, there are gaps between the bricks called weep holes, depending on the builder. Some are every third brick, some are every second brick, and some are every brick. The weep holes are so your house can breathe in the summer heat and allow moisture to escape to prevent mold and allow fresh air to get into the inside walls. The problem is that it is an open door for all kinds of small critters such as crickets, mice, big tree roaches and even small roaches, snakes, lizards, bees, wasps, earwigs, beetles, pill bugs, spiders, and whatever can fit through that hole in the brick. Now you cannot seal them with caulking or silicone because your house needs to breathe. However, you can purchase a weep hole cover that will keep out most pests and still allow air in and allow your house to breathe. They make some plastic ones; I, however, have found them to not lock in tight enough and the weed blower can blow them out if the builder leaves the gap too big. My company uses a stainless steel one that grabs the brick and does not come out, regardless of the gap size. They also do not cause rust stains on the brick. This is my number one tip for keeping bug outside without using poison and a good solution to keep the pest away and not pay an exterminator.

Mosquitoes Are Next. Here is a little trivia for you! That nice little fogging truck that your city or county uses to spray for mosquitoes seems to do a great job, or so we are led to believe.

I was at my continuing Education class years ago when a Bayer Rep named Ed asked the three hundred professionals in the room what the kill rate for a fogging truck was. For five minutes, people shouted out answers as he stood in front of the room holding up one finger. Finally, the crowd got quiet as he stood holding up that finger, and he told us the answer was that a fogging truck has a 1% kill rate. That is a horrible kill rate, but companies get paid hundreds of dollars per hour to run that truck around with a 1% kill rate.

I caught Ed at the break and asked why and how to do things that actually work, and it led me to a great awakening on mosquito treatments. He taught me about using a larvicide to kill them before they become flyers. Ed has had malaria twice in his quest to combat the little vampires we call mosquitoes. Standing water is key to mosquito control in things like cans, pots, and old tires. Anything that can hold water is a breeding site for mosquitoes. Garlic Oil is a great product to use on mosquitoes; it kills and repels for about thirty days and is non-toxic to your family, unlike things like control44 and malathion and bifenthrin. Mix three to six ounces in a gallon of water in a pump sprayer and spray before ten am or after five pm. Mosquitoes hide from the sun during the day on the bottom leaves of the bushes around the house and anywhere shade is found. That is where to spray the Garlic oil. In2Care mosquito traps are another safe and effective treatment without harsh chemicals. See a video on our website, www.thegreenbugman.com. They are cool because the female goes into the trap that is filled with water to lay eggs, and when she does, she touches a mesh screen and contracts a fungicide type virus that kills her in about three days. But here is the cool part as she is flying around laying eggs in standing water, she is contaminating the water for all other female that lay eggs in that water for the next three days. It is the best thing we have ever found for mosquitoes by far. Our web site has a great video that explains how they work at www.thegreenbugman.com

Winning the War Against Spiders

Research suggest that you are never more than ten feet away from a spider. There are over 30,000 different species of spider that are known. And yes, it is true that the female spider will eat the male spider before, during or after copulation. Spider webs are five times stronger than the same strand of steel of the same thickness. Did you know spiders have blue blood? The spider web is liquid. When the silk encounters the air it hardens allowing them to create and build their webs. Enough food there to satisfy you useless trivia for a while.

Not all spiders will bite. They just want to be left alone doing spidery stuff like eating other bugs. I have known people who have pet tarantulas in their house and take them out and play with them. Neither has ever been bitten. When spiders feel threatened, they bite. Let's be real; you are too big to be food for them. They kill for food or attack when they feel threatened.

My pest control company is in south Texas, and we have lots of spiders We have a few black widows, but I see a lot of brown widows. You will hardly ever see them unless you are hunting them. You will see the egg sack before you find the momma. It Is a small fuzz ball with small little spikes on it. The egg sacks will be around window trim door frames and even doorknobs that see little use, or. under the eaves of the house away from the weather as much as possible. She will be hiding very close to her precious egg sacks within a few inches. There are labeled pictures at the end of this little book.

Now let's talk spraying for spiders!

All your insecticides are microencapsulated. Which means they are made to be sticky, and the bug feels it licks it off and ingests the poison and dies. Ever get gum on your shoe? Microencapsulation works the same way. Got to get that sticky stuff off now! That is because most bugs are social insects meaning that they groom them selves and others they are in contact with them. Spiders are considered a non-social insect meaning they don't groom themselves. So, a roach walks through a sprayed area he cleans himself and will die. Now his friend the spider can walk through the same sticky product and not die. Since they are nonsocial, they do not groom them selves and ignore what they have stepped in. So, you must spray the body of the spider to kill him with a spray,

Here is a potential risk when spraying a spider up high say on the ceiling or wall above your head. Remember what goes up must come down, so when you spray over your head the drift from the spray comes down and can get in your eyes and on even in your mouth and eyes. I had a customer who did just that and was blinded by the spray that fell in his eyes, he then missed a step and ended up breaking his ankle. If using bugs spray it is some for the spider and some for you, which is no good for you, but it will kill that little spider. Try using window cleaner instead of bug spray. It will kill the spider and be much safer for you.

While we are talking about sprays, I must mention my secret weapon Web Out. It will kill and repel spider and it is all natural it is a 25B product and that means that it does not require a chemical label at ALL! 25B products are label exempt simply because they have been found to have no adverse reactions in humans. Even things like 409 and other household cleaners have chemical labels as they can cause harm to humans. Web Out is Thyme oil, Citronella oil and Lemongrass oils and comes ready to spray right out of the bottle. For spiders spray around all windows and entry points like around door frames, chimneys, under the eaves of the house and around the outside of all doors and all exterior lights. Web Out I heavy oil base and great to use on say a boat dock and on a boat and it will not harm the aquatic life or the boat and holds up well it the weather.

In our company we have tried it on many pests and found really great results. For flies, wasps, American roaches even carpenter bees.

We tested it on a dumpster in a commercial property which was covered with flies. The flies left quickly. Wasp will always return to the place where they were born to have babies. We found that if we knocked the nest down and sprayed Web Out where the nest was, they would noy come back to that area. Carpenter bees are really quite amazing they drill a small hole and it forms a perfect ninety degree angle then pack it with eggs and the eggs will hatch in reverse order. A customer had been battling them drilling hole in his two Palapa's in his back yard. We sprayed Web Out and, no more carpenter bees! We service some old building, and in one we kept sticky traps in the mechanical rooms, and we would get six to ten American roaches every month by the drains. After spraying Web Out each month and we went done to one roach per month. Web Out does a lot but, it is the BOMB for spiders.

The Brown Widow **The Brown Widow egg sack!**

Not the best method! (don't try this at home)

Winning the War with Ants!

I have been in pest control for twenty years and have operated my own company for twelve years. The following are the most common mistakes that people have told me they have made and the correct way to get real results.

You go out and purchase ant killer from the store one that kills the queen. Then days later you have more ants! Not true, these are the same ant colony because secondary reproductive ants are doing what they were born to do, KEEP THE COLONY ALIVE!

The most common thing I have heard from people about ants is that they treat the ants and a few days later a new mound will appear to replace the old one a few feet away from where they just treated days before. It is not that the produc does not work after all they market that it kills the queen dead. And it does just that, it kills the queen just as promised. First you must understand your enemy when you go into battle. Ant colonies have one major goal and that is that the colony comes first and must survive at all costs. Marketing ant killers is all about killing the queen, one says that it kills the queen dead, and they want you to believe that killing the queen is all that matters. It is very misleading, as is most marketing. The truth is killing the queen is easy, as she is nothing more than a fat egg laying machine and is an easy target. What they don't tell you is this.

The colony has a lot of secondary reproductive ants that are just waiting to take the queens place should she die. Remember rule number one, the colony must survive.

These secondary reproductive ants are living all around the edge of the colony, well away from the queen and her throne. So, when you walk up to the mound to treat it and battle them to kill their queen you vibrate the ground, and they know you are coming and their instincts kick in. When an attack comes, the secondary reproductive ants sound the alarm and haul ass away from the mound and the queen. You apply your chosen ant killer as planned but the secondary reproductive ant's escape!

You apply your poison to the center where the queen is, and it works! She dies and the mound dries up and dies and you see bunches of dead ants the next day and you celebrate for you have won! Only you go back out days later and there is a new mound formed up waiting for you. The problem is those secondary reproductive ants have done what they were made to do in case the queen dies. Rebuild and keep the colony alive. Remember; "kills the queen dead" It has done just that; given you a victory, even though it is short lived. But guess what you can just get more ant killer.

The Solution!

The Way to achieve victory and use a lot less any killer is this; When you walk up to attack the ant mound you vibrate the ground with your footsteps. Does this mean you have to wear a ninja costume and crawl up to the mound is stealth mode? Al though that might be fun, to be the killer ant ninja. It is not necessary. When you begin treating start about a foot away from the edge of the mound or more if it is a large mound. This puts your any killer in front of the escape path of those little secondary reproductive. Now you can attack the actual visible mound and kill that queen dead. And the secondary reproductive ants will die and so shall the colony.

In the world of pest control it is about recurring services- teaching you to rely on the pest control company or a product to keep the bugs away. Hence the quarterly service contract. The bug guy coming on a regular basis to keep your home pests controlled. Or getting you back to buy your favorite ant killer as soon as possible.

Next, we move on to the nomadic ant. These ants do not build a mound like the fire ant does. You will se a few ants or even a trail of ants but not find a mound like our fire ant friends. What do most people do is grab the bug spray and kill those little invaders. These little guys do not mound but rather build little nests under things around the outside of your house. Place like under your patio, under your AC Pad, under planters and pots, under walking pavers or anything else they can get under. Remember, ants are all about the colony surviving. One colony will have multiple nests around your property. You kill one but the colony survives. Again, most people spray these ants and that is the mistake. You kill a nest but when ant colonies are attacked the divide and rebuild like the fire ant moving to a new spot in the yard. You get some relief with your trusty spray and are ready next time they come back, and they will. These types of ants are very chemical sensitive, so sprays do not work well on getting then to stay away. Bait and wait is what it is called in pest control. Bait and wait for it to do its magic. Ants operate through pheromone or scent that way the other ants can find food or water.

Let's move on to the American roach. They are much larger than the German roach and can be as large as two inches. Unlike the German roach, they are flyers. German roaches have wings and cannot fly. However, their cousin, the Asian roach, can fly. American roaches are sometimes called water bugs. Here is a fun fact about them; they can hold their breath for around forty minutes, so if you flush one down the toilet, make sure it is dead, or it will be back. They are roach that loves bacteria and are seen in and around sewers.

These guys enter your home through your weep holes in a brick home. Weep hole covers can reduce their entry point and keep them outside. In a non-brick home, look under all your sinks where the pipes come through the wall. Even if it has nice plastic covers, slide them forward and seal any holes around the pipes. Baits work well, and sticky traps also work well against them. They see the folded trap as a hiding spot and run in and get stuck for the rest of their life. Garage doors are a bug entry point for them as well. Again, sticky traps on both sides of the garage door work well, are cheap and safe and work 24/7. They do not infest and take over like the German roach; they are considered an occasional invader, prefer outdoors, and will invade when it gets too dry or just before a storm. And yes, there are more types of roaches, like the brown-banded and many others. It doesn't make much difference what species they are. They must be eliminated or controlled.

Cockroaches creep out many people, and we teach you more than you probably want to know about these little bugs. We will cover the most common roaches in this book.

Let's talk about with the German roach. They are small roach that is about 1.1 to 1.6 CM long. They vary in color from tan to almost black, with two dark, roughly parallel streaks on the pronotum running from the head to the base of the wings. They can carry salmonellosis (food poisoning). Also known to carry typhoid fever, cholera, gastroenteritis, dysentery, listeriosis, giardia and E. coli. They pass these to humans by walking over food or prep areas, utensils, and other things.

Here is a fact about the German roach, they only travel three feet in their life. Well, how the hell do they get into your home? They are transported by us without even realizing it. I recently opened a bag of dog food from the grocery store, and several German roaches were hiding in the folded edge of the bag. They ran and fell out into my kitchen when I opened the bag. Had the same issue with a bag of cat food. They love cardboard boxes inside the corrugated raised areas, and they feel safe when their feet and antennas touch something. They can hitch a ride wherever the box goes, from a warehouse to the grocery store and hen to your house. Electronics can carry German Roaches just like the dog food bags. Small appliances or food can be put in the freezer overnight to kill and oaches living in a laptop or a coffeemaker and even a small microwave.

Roaches spend eighty-five percent of their day tucked in somewhere and hiding. They like to feel something touching their feet and antennas to feel safe. They ome out of their hiding spot to eat, drink and poop. Remember the three-foot rule. They reproduce faster than any cockroach, growing from egg to reproductive adult in roughly fifty to sixty days. Each egg sack carries thirty to forty babies and an lay thousands of baby roaches yearly. They can produce one egg sack or capsule after mating and can produce two in one week. They produce an egg sack about every six weeks and with thirty to forty in each sack, do the math, and it is crazy how many they can produce in a year, and that is just one roach. Multiple the number of ten roaches having forty babies every six weeks, ten females can produce four hundred babies in just six weeks.

So, spraying bug spray on the one you saw when you turned on the light will kill that one, but what about the egg sack full of babies? No spray will penetrate that egg sack. You must wait till they hatch to kill them. What about food sources? Well, first off, roaches are cannibals in that each will eat the bodies of other roaches for food. So just keeping the crumbs cleaned up will not keep them from eating, and they can live off just water for 3thirty days.

German roaches do not like temperature change; they like a constant temperature. That is why they like electronic equipment and a nice warm temperature. Places like the fridge motor where they can find water, food, or dust. They eat dust for protein content, and there is always dust under the fridge. Television stereos, game consoles, coffee makers and even microwaves. This is why you should beware of used appliances or equipment. If you get a used laptop, it would be good to freeze it and kill any bugs that may be hiding inside. Put it in a large zip lock bag, seal it and twelve hours in a freezer will kill all the bugs. If it won't fit in the freezer or you don't want to try that, simply put a small amount of roach bait in it. Simply apply some roach bait around the motor or electrical part to be safe.

If you hire a company, make sure they follow up after the first treatment about two weeks out to get the babies from the egg sack that hatch. Remember, no spray or bait can penetrate the egg sack. If trying to do it yourself, rely on putting out roach bait over the spray.

Wasps and Bees Are Up Next.

Wasps will always return to where they were born to have babies. Trivia fact time; a dirt dobber packs egg on top of each other in their little dirt nests, and they have in reverse order. Last egg hatches first and then the next one after tha all the way to the first egg in the nest. Dirt dobbers are a passive wasp and you have really try hard to get stung. Most wasps and bees are not aggressive; however, when you fling yours hands and arms in the air they see it as an act of aggression. Hands down and remain still and they will buzz around you and then leave if you remain calm and are non-threatening. I freak customers out when walk up to a bee swarm. When you get to close, they will buzz you and it is a warning to back off. When I fight them, I always use a bee suit so I don't get stung. Again, most wasps are the same except yellow jackets, They are just plain mean and have stung me when I get to close and do not even see them. Carpenter bees are very mild-mannered, and you must work hard or just be stupid for them to sting. Again, hands down and remain calm. Hands down is the key to not getting stung.

Bees will swarm looking for a new home in the spring and if they pose a threat to kids and pets the best way to kill them is Dawn soap in a bucket and pour it over them. Remember we need bees to pollinate our flowers and crops. Contact a beekeeper and have them relocated.

Let's Talk Fleas Now.

How to recognize flea bite; knees down is normally flea bite. Fleas tend to come from the ground and bite as quick as possible so knees down is normally flea bite as opposed to bed bug bites, which are normally from the aim pits up. With fleas you must treat the host. For dogs and cats there is a product call cedarcide and it is a natural flea killer and repellent as well as garlic oil. Meet a dairy farmer in Belize when I lived there. He put garlic oil on the legs and tail of his cattle to keep fleas off and not use chemicals.

For the yard treatment there is nothing natural that works except Cedar oil and Garlic oil. They can be controlled by using over the counter ant killer. Bifenthrin is the active ingredient and water will not kill it like most product, it is a man-made synthetic chemical and very effective and at a decent cost compared to the natural oils if you have a large yard. If you have fleas inside of the house, you need to vacuum all carpet and rugs and wait eight hours before you treat. The vibration from the vacuum will cause the eggs to hatch. No chemical can penetrate the egg shell so you must force them to hatch. I get at least two calls every year from people who come back from vacation and get attacked by fleas. They eggs were laying in wait and when they returned home and vibrated the floors it creates a flea explosion and those are hunger little dudes are out for blood. Side note: fleas can live through a vacuum cleaner so if you try to vacuum up the lives ones just to get relief, TAKE the vacuum bag or canister outside of the house when you empty it or risk them escaping into the house.

Brief Look at Bed Bugs.

Bed bugs from a pest company's perspective can be a nightmare, because it requires the customer to take things to the cleaners or wash massive amounts of clothes. I explain that clothes and shoes and even drapes that are close to the floor need to be cleaned or washed in hot water and it's best to use a laundry matt so you can make one trip. Once they are taken out o the bag at the laundry mat in and cleaned do not put them in the same bag you brought them to the laundry mat. Put them in a clean bag for the trip home . Our company quit doing bed bug treatment after a few customer did not do their part correctly and needed extra treatments.

Now Let's Talk Termites and the Wood Destroying Insects.

The list includes all three types of termites, wood boring beetles, carpenter bees and ants.

Termites come is three types Eastern subterranean, Drywood and Formosan Subterranean. Drywood live inside the wood as opposed to the Formosan and Eastern subterranean who live in the dirt and come from the ground up to eat on your house. Termites will eat anything that contains Cellulose such as books , wall paper pictures of Grandma on the wall. I actually found a picture of a grandma who had been ravage by the termites. If it has paper, they will destroy it. Termites are very active before swarm time. Eastern subterranean swarm in early spring as early as February but usually start in March or April. They swarm mid-morning from ten am up through two pm. Formosan swarm in late May and into June. They swarm at nightfall, about the time the streetlights come on, and in large numbers. All swarming termites are attracted to light. Eastern look for sunlight to get outside and you find dead ones in the windows. Drywoods swarm in summer like the Formosan, but in much smaller numbers. All termites have a mating time when they swarm out the Elates AKA swarmer-s, they produce termites with wings and their job which is to fly out away from the existing colony to start another colony. For the subterranean termites you need treat everywhere the house touches the ground. For Drywood you must treat the wood itself and the most common way is through a fumigation where they cover the house with a huge tent and pump Vikane gas into the tent to kill the termites. Sometimes with Dry wood termites it is easier and cheaper to just remove and replace the damaged wood. The trick is to go three to six feet past where they are and replace the wood. Drywood colonies are small, and they are slow eaters compared to the subterranean. An Eastern subterranean colony eats a board foot of lumber in about two month which is about a six foot 2 x 4 in 2 months versus a Formosan colony which can eat that same six foot 2 x 4 in just *two* days. They are little monster due to the size of the colonies. They are not something to mess with and put off.

Now let's get into the carpenter bees and ants. They destroy wood but not near as bad as termites. Carpenter bees drill a perfect ninety degree hole into the wood and pack it full of eggs and like the wasp they hatch in reverse order and will have there eggs in the same area where they were born. The best natural remedy for them is to spray the holes with Web Out a natural product that is 25 B exempt from a chemical label. Spraying it won't penetrate the egg sack, but it will make them leave the area.

Carpenter ants are a little trickier simply because there is no mound to attack. They usually come from a tree near the house. They do not create a trail like most ants. There will be one or two ants and they are larger than the average ants. You just must watch and see where they disappear into the house and the nest will not be far away, most times you can treat the tree where you find them. There is a special ant bait for them that you just sprinkle in their path and they will take it back to the colony and die. I have yet to find a natural product that works on them.

Now onto the wood destroying beetles, Powder post beetles are the most common and they are similar to drywood in the they live inside the wood, so it requires replacing the damaged wood or tenting, they are slow movers and are pretty low on the damage causing scale. I have found them in furniture quite a few times and in that case the best method is to put the piece in a freezer for twenty-four hours and they will die. Biggest problem is finding a freezer. I found a deer processor and when it is not deer season his was available. Or there are companies that fumigate and tent containers and furniture. Here is a scale on damage causing insects.

1 The worst one is the Formosan termite A six foot 2 x4 in just two days. These guys always eat with the grain of the wood and never eat the ring of the tree or wood.

2. The Eastern Subterranean Eat that same 2 x 4 in two months versus two days and always with grain, These types of termite and the Formosan take dirt with them everywhere they go so if you find wood with dirt inside it you have termites.

3. Drywood are next on the damage scale. The colonies are smaller and localized in the wood itself. They eat cross grain and will deposit Frass (termite poop) out from a small hole. It is very hard and rubbing it in between your fingers it will not crumble, and rub it long enough it will blister your hand it is so hard and it looks like coffee grounds.

4. Next on the scale are carpenter ants. They tend to like softer wood that has gotten went usually around windows and door frames. The do not eat the wood but they Shred-it and build a nest inside the wood.

5. Next one on the scale is Carpenter bees. They drill small hole around 3/8 of and inch wide into eves or soffits on a house-or in Palapas. But when they find an area, they come in large numbers. If you see small random holes is eaves and soffit, they are from carpenter bees.

6. The wood boring beetles are last on the damage scale. They move really slow and their numbers are small. You will find a small pile or very fine wood dust from their little kick holes.

I recommend calling a company to deal with termite problems They are not something you can treat with a one-gallon pump sprayer. Most three bedroom house require fifty or more gallons of termiticide to treat the house effectively and a pretty stronger amount of pressure. We have a sixty-gallon tank with a pump cranking out sixty to eighty PSI. Like a pressure washer. You want to saturate the dirt with the chemical and the chemical will stay in the dirt for around ten to fifteen years, It will also contaminate a well and will kill aquatic life if it gets in the water or pond. Another reason to call a professional. I will say this I was taught in bug school to scare the heel out of people when it came to termite damage. Fear is a huge way to motivate customers to buy from you and not shop around like you would normally do. Do not give into fear-based salespeople.

Last Pest on The List is Snakes.

I am in pest control and don' t like these guys at all. When I do crawl space termite work and inspections I carry a machete with me in case I find a snake. It's him or me and one of is going to die!

I once went to a house for a termite inspection and before I went into the crawl space I took my machete from the truck as always. The customer asked why I needed that I explained about my dislike of snakes, and she told me her son had a six foot Boas snake that is loose in or under the house. I told her it was me or the snake, and she asked me to leave if I was going to kill the snake. I left fast. I could not imagine being in a small crawl space and finding a six-foot snake! My worst nightmare possible. I dove away that day!

Snake will follow a food source which they love rats and mice. I was at a house for mice when I opened the closet where I found about six mice living with a snake all curled up in the corner. He was a small one, but he went to snake heaven that day along with his mouse friends. Snakes get into your house usually through the weep holes chasing mice and bugs. Weep hole covers will stop them along with snake-away. Snake Away is Naphthalene and Sulfur and creates a sense of fear and confusion in the snake. It will last about thirty days as long as no rain gets it.

This book was made to educate you as a homeowner so you can make a good decision not based on fear but one you feel good about. Pest control companies have a place and a purpose if you want or need one; however, it is doable to handle most pests on your own if you have the desire to do so.

Lastly will be a list of resources for you to use and everything can be found on the internet now. Thank God for You Tube. I am a car guy and do most of my own repairs and I custom build my own cars. When I was younger (I'm sixty two at this time) we did not have the internet yet, and you had to pick up a land line. (a phone attached to the wall) and call and friend or get put the big yellow book to find answers.

1. ECO PCO AR-x -A great multi-purpose safe bug spray in an aerosol can
2. Web Out - An awesome natural product made for spiders but works on a wide variety of critters. Great for boat slips and boat docks.
3. Cedarcide and Wondercide - Both are made from Cedar oil and are awesome for fleas and mosquitoes.
4. 4. Garlic Oil, AKA mosquito barrier - safe and effective for mosquitoes and fleas.
5. Sticky Traps - small foldable glue boards that work 24 / 7 and catch any-thing that goes inside them, roaches, mice, crickets and lots of other critters. They also made bigger ones in a plastic tray for larger rodents. If any- thing like say a cat or dog or a child's small hand gets stuck in these traps simply use some vegetable oil or things like WD-40 to loosen the glue up.
6. Ant bait and roach bait can be bought at places like Home Depot or Amazon. Look for the active ingredients of either Imidacloprid or Friponil. They are the best and it just takes a small amount about half the size of a pencil eraser.
7. Stuff you can't buy without a pest license like termiticide and In2Care packets for mosquitoes.
8. Snake away is best thing to repel snakes and it can be bought at all the big box stores and Amazon.

Our web site is www.thegreenbugman.com and we have all the products talked about in this book. If you have any questions about bugs or stuff in the book or anything bug related email us at bugmantom@gmail.com.

The Green Bug Man was founded on finding the safest and greenest way to control and effectively stop the insects. We even use all natural products in our home and office that are non-toxic and cost effective.

In Closing, I will say this 'fear sells and a knowledgeable customer is your best customer. Go forth with an understanding that there are safe and effective things you can do yourself if you desire. If not, this may help you to get a quality type service provider to solve your problem.

Table of Contents

Acknowledgements:

I want to thanks my Mother and my sister Kandy for providing me with another way of looking at problems and the freedom to think outside the box.

Made in United States
Troutdale, OR
11/03/2026

56852702R00015